Delphite & Jadite

A Pocket Guide

Joe Keller & David Ross

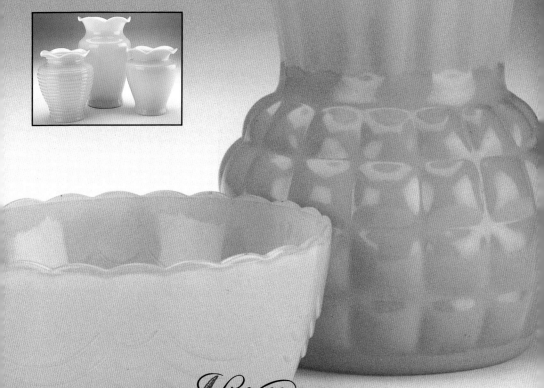

Schiffer Publishing Ltd

4880 Lower Valley Road, Atglen, PA 19310 USA

Copyright © 2002 by Joe Keller and David Ross
Library of Congress Control Number: 2002108364

All rights reserved. No part of this work may be reproduced or used in any form or by any means—graphic, electronic, or mechanical, including photocopying or information storage and retrieval systems—without written permission from the copyright holder.
"Schiffer," "Schiffer Publishing Ltd. & Design," and the "Design of pen and ink well" are registered trademarks of Schiffer Publishing Ltd.

Designed by Bonnie M. Hensley
Cover design by Bruce M. Waters
Type set in Humanst521 Cn BT/Humanst521 BT

ISBN: 0-7643-1640-0
Printed in China
1 2 3 4

Published by Schiffer Publishing Ltd.
4880 Lower Valley Road
Atglen, PA 19310
Phone: (610) 593-1777; Fax: (610) 593-2002
E-mail: Schifferbk@aol.com
Please visit our web site catalog at **www.schifferbooks.com**
We are always looking for people to write books on new and related subjects. If you have an idea for a book please contact us at the above address.

This book may be purchased from the publisher.
Include $3.95 for shipping.
Please try your bookstore first.
You may write for a free catalog.

In Europe, Schiffer books are distributed by
Bushwood Books
6 Marksbury Ave.
Kew Gardens
Surrey TW9 4JF England
Phone: 44 (0) 20 8392-8585; Fax: 44 (0) 20 8392-9876
E-mail: Bushwd@aol.com
Free postage in the U.K., Europe; air mail at cost.

Contents

Introduction and Acknowledgments 4

Delphite Kitchenware and Dinnerware 5
    Delphite Kitchenware 7
    Azurite Charm 38
    Azurite Swirl 41
    Turquoise Blue 43
    Fire-King Kitchenware and Miscellaneous Delphite 49

Jadite Kitchenware and Dinnerware 62
    Jadite Kitchenware 64
    Alice 127
    Charm 128
    Jane Ray 130
    McKee Laurel 136
    1700 Line 144
    Restaurantware 141
    Sheaves of Wheat 148
    Shell 150
    Swirl 154
    Martha Stewart 156

Introduction & Acknowledgments

Jadite and Delphite are the most frequently collected Depression Era kitchen glassware. This book is intended both as a beginners' guide and a portable resource. The Jadite section is obviously less comprehensive than our *Jadite, An Information and Price Guide,* which includes over 500 photos. This book contains items that are most frequently found at antique markets and Depression Glass shows. We have also tried to include realistic pricing information that may help in assessing the value of items.

This book has been created by using photos from many collector's collections. Special thanks to the following individuals for allowing us to use their collections and resources: Molly Allen, Jane and Jerry Bohlen, Ruthie and Winfield Foley, Randy and Dixie Hardesty, Dennis English, Cookie Katz, Joe Keller Sr., Anna and Joe Thomas, Suzanne Weatherman, and Dalen Whitt.

Delphite Kitchenware & Dinnerware

Long before the recent Jadite craze, opaque blue kitchenware, known as "delphite," "delfite," and "chaline," was passionately collected. While it originally sold for the same amount as its green counterpart, today on the secondary market it is much more expensive and much harder to find than Jadite.

Jeannette hat advertising "Delfite."

Turquoise Blue dinnerware.

First introduced by the McKee Glass Co. in the late 1920s, delphite was produced in much smaller quantity than Jadite. Several distinct colors exist. Products by McKee and Jeannette do not match in color. This is further complicated by McKee producing both delphite and chaline. Chaline is the lighter more vibrant color. While there are many collectors for both colors, delphite appears to be the preferred color in today's market.

Anchor Hocking produced several dinnerware lines and some kitchenware in opaque blues under their Fire-King lines in the 1950s and 1960s. They produced two distinct colors, Turquoise Blue and Azurite. Azurite is an extremely pale blue that appears almost white. Turquoise Blue is a richer more vibrant blue. These Fire-King lines are still affordable in comparison to Fire-King's Jadite lines.

Delphite Kitchenware

Jeannette delphite kitchenware.

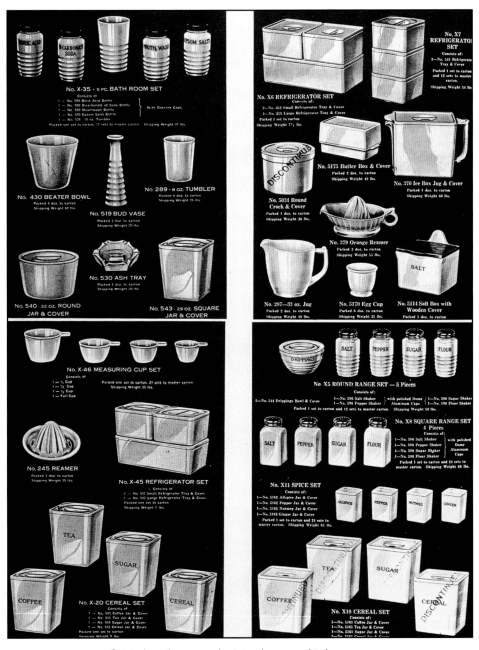

Original catalog pages depicting Jeannette kitchenware.

Jeannette delphite canisters. Sugar, $400-450; tea, $275-300; coffee, $375-400.

Jeannette 29 oz. canisters. $325-350 each. Flour (not pictured), $375-400; coffee (not pictured), $325-350.

McKee delphite 48 oz. canisters. $650-750 each.
McKee 48 oz. Chaline canisters. $400-450 each.

McKee Chaline canisters. Small, $45-50; medium, $125-150; large, $125-150.

Jeannette shakers. Pepper, $80-90; salt, $80-90; flour, $150-175; sugar, $150-175.

Jeannette shakers: Flour, $125-150; paprika, $200-225; salt, $75-85; pepper, $75-85; sugar (not shown), $125-150.

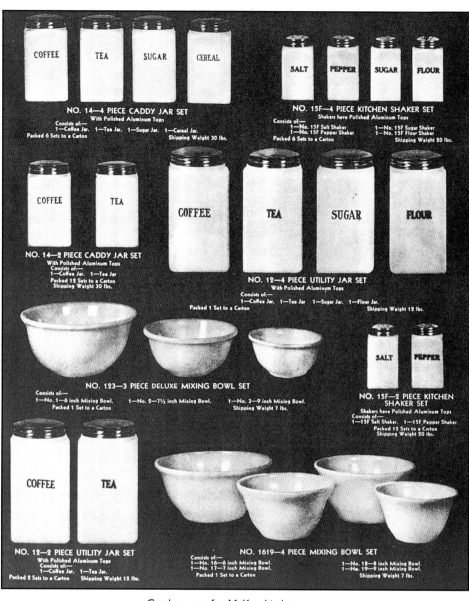

Catalog page for McKee kitchenware.

McKee small box delphite shakers. Salt, $175-200; pepper, $175-200; flour, $275-300; sugar (not pictured), $275-300.

McKee Chaline shakers. Salt, $75-100. pepper, $75-100; sugar, $125-150; flour, $125-150.

McKee Roman Arch delphite shakers. Salt, $225-250; pepper, $225-250. Flour and sugar (not pictured), $500-550 each.

Delphite ginger, $500-550; jadite, $350-400.

Opaque blue pepper shaker with embossed letters. $200-225.

Original boxed set for Cook's Coffee Company. Drippings, $275-300; matches, $150-175; original box, $25-35.

Delphite shakers, $20-25 pair.

Jeannette refrigerator jars. Round, $75-85; 4"x 4", $45-50; 4" x 8", $75-85.

Jeannette butter, $250-300.

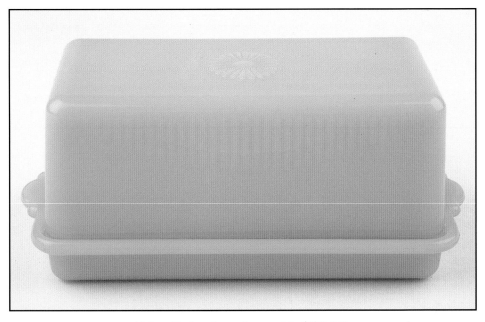

McKee Chaline cheese/butter dish, $200-225.

McKee advertisement for kitchenware.

McKee Chaline 6" mixing bowl, $90-100; 7-1/2" (not pictured), $100-125; 9" (not pictured), $125-150.

Jeannette delphite mixing bowls. 6", $50-60; 7", $60-70; 8", $70-80; 9", $100-110.

Delphite mixer bowl. Scarce, $125-150.

Jeannette delphite measuring cups. One-quarter cup, $40-45; one-third cup, $50-60; one-half cup, $50-60; one-cup with spout, $80-90. Original box, $40-50.

McKee delphite measuring pitcher, $200-225.

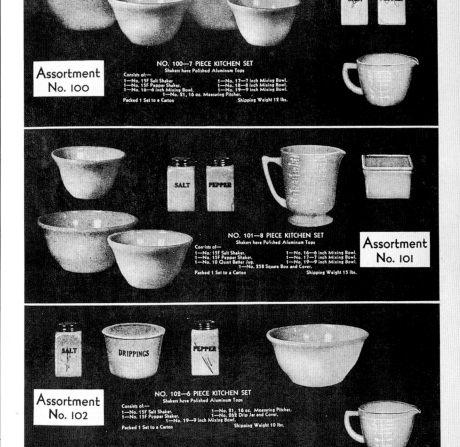

McKee catalog page.

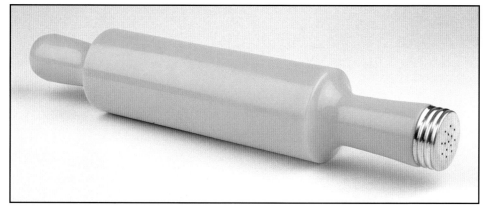

McKee chaline rolling pin, $1500-2000.

Jeannette delphite batter bowl with beater, $100-125.

Jeannette batter bowl with electric motor, $75-100.

McKee Sunkist reamer, $125-150.

McKee Chaline vases. 8" ribbed, $100-125; 12" plain, $150-175; 8" plain, $100-125.

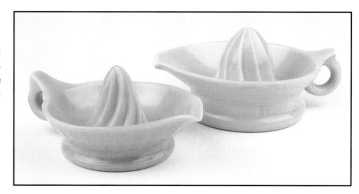

Delphite reamers. Left: Jeannette lemon, $90-100. Right: McKee orange, $500-700.

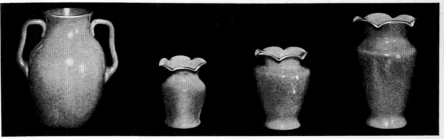
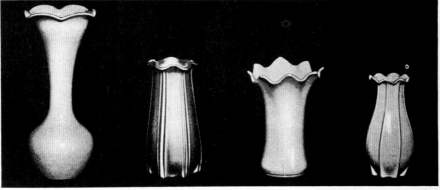

Catalog page depicting McKee vases.

McKee Chaline nude vase, $250-300.

McKee comport, $60-70.

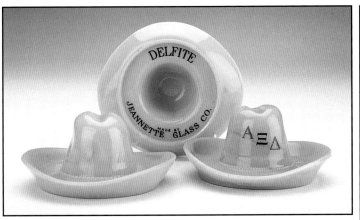
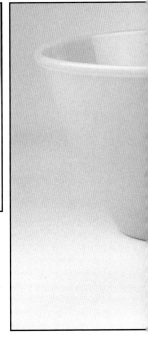

Jeannette hats. Plain, $15-20; advertising "delfite" hat, $175-200; fraternity souvenir hat, $20-25.

Fire-King splash proof mixing bowls. One-quart, $35-40; two-quart, $30-35; three-quart, $30-35.

Plain hat, $15-20; round ashtray, $50-60; hexagonal ashtray, $30-35; Souvenir hat, $25-30.

TURQUOISE-BLUE MIXING BOWL SETS

NEW SWEDISH MODERN SHAPE

B500/39—4 Pce. Mixing Bowl Set
  Each Set in Gift Carton, 4 Sets to Shipping Carton — 31½ lbs.
  COMPOSITION: One 1 Pt. Mixing Bowl
    One 1 Qt. Mixing Bowl
    One 2 Qt. Mixing Bowl
    One 3 Qt. Mixing Bowl

B500/40—3 Pce. Mixing Bowl Set
  Each Set in Gift Carton, 4 Sets to Shipping Carton — 19 lbs.
  COMPOSITION: One 1 Pt. Mixing Bowl
    One 1 Qt. Mixing Bowl
    One 2 Qt. Mixing Bowl

SPLASH-PROOF SHAPE

B300/213—3 Pce. Mixing Bowl Set
  2 Dozen Sets Bulk Packed in 4 Cartons — 127 lbs.

B300/214—3 Pce. Mixing Bowl Set
  Each Set Nested and Packed in an Individual
  Cell — 8 Sets to Shipping Carton — 40 lbs.

The Sets listed above consist of one each of the
B366, B367 and B368 Bowls.

OPEN STOCK PACKING
B366—1 Qt. Mixing Bowl   2 doz. — 28 lbs.
B367—2 Qt. Mixing Bowl   2 doz. — 43 lbs.
B368—3 Qt. Mixing Bowl   1 doz. — 28 lbs.

HEAT-PROOF

Mixing Bowls
in Assorted
Colors

See Jade-ite and
Ivory Mixing Bowls
on Pages 28 & 29.

HEAT-PROOF

Colors
Will Blend
With Any
Kitchen

Crystal Mixing Bowls
are on
Page 32.

W300/149—4 Pce. Mixing Bowl Set
  Each Set in Gift Carton, 6 Sets to Shipping Carton — 35 lbs.
  COMPOSITION: One 4⅞" Mixing Bowl—Green
    One 6" Mixing Bowl—Blue
    One 7¼" Mixing Bowl—Yellow
    One 8⅜" Mixing Bowl—Red
    (Available only in Set Packing)

Decorated Mixing Bowls on Page 26.

Fire-King catalog depicting mixing bowls.

Swedish Modern mixing bowls. One-pint, $30-35; one-quart, $50-60; two-quart, $50-60; three-quart, $45-50.

Fire-King batter pitcher. Very rare in Turquoise Blue. $400-500.

Fire-King ice tub, $60-75.

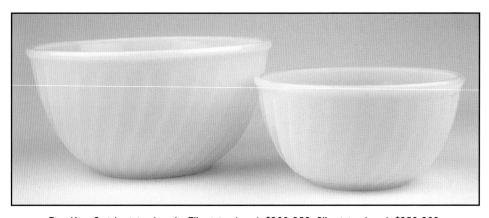

Fire-King Swirl mixing bowls. 7" mixing bowl, $200-250; 9" mixing bowl, $250-300.

Ribbed mixing bowls. Jadite, $40-45; azurite, $50-60.

Fire-King rare blue Art Deco vases, $200-250 each.

Fire-King three-footed planter, $30-35; fired-on bulbous vase, $20-25.

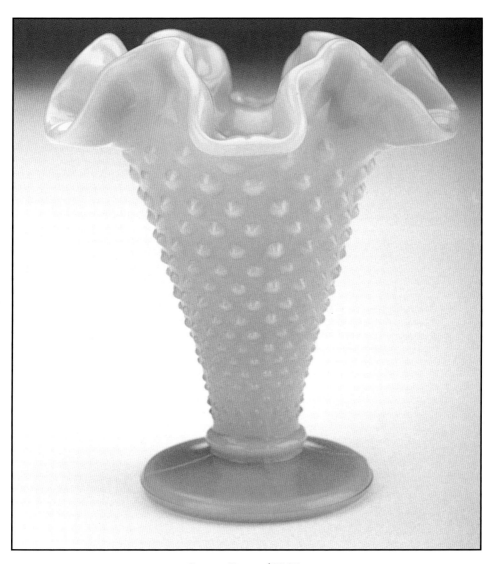

Fenton 4" vase, $20-25.

Azurite Charm

Charm Azurite dinnerware.

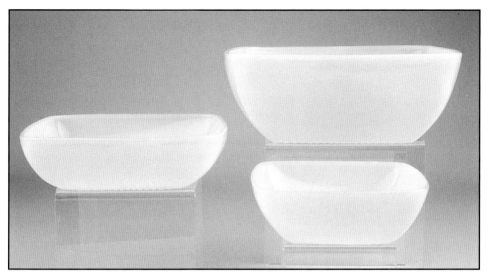

Charm bowls. 4-3/4" dessert, $8-10; 6" soup, $18-20; 7-3/8" salad bowl, $30-35.

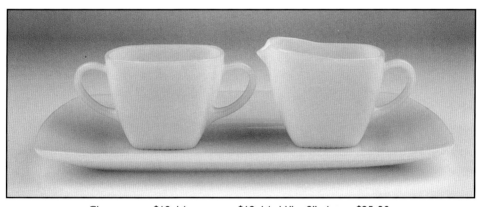

Charm sugar, $12-14; creamer, $12-14; 11" x 8" platter, $25-30.

Opposite page, bottom
Charm plates. 6-5/8" salad, $12-15; 8-3/8" lunch, $10-12; 9-1/4" dinner, $18-20.

Charm cup/saucer, $5-7.

Charm novelty plate, $20-25.

Azurite Swirl

Swirl dinnerware.

Swirl plates. 9" dinner, $10-12; 7-3/8" salad plate, $5-8; platter, $18-20.

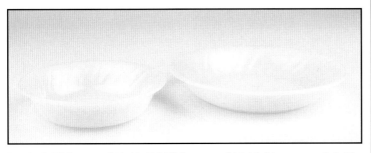

Bowls. 5-7/8" cereal, $10-12; 7-5/8" flat soup, $12-15. Not pictured: 8-1/4" vegetable, $20-22; 4-7/8" dessert, $8-10; 9" flanged soup, $75-100.

Swirl creamer, $10-12; covered sugar, $10-12.

Turquoise Blue

Turquoise Blue plates. 10" dinner, $25-30; 9" lunch/dinner, $8-10; 7-1/4" salad, $10-12; 6" bread plate, $15-20.

Turquoise Blue bowls. 4-1/2" berry, $8-10; 6-1/2" soup, $20-22; 5" thin cereal, $35-40.

Turquoise Blue sugar, $8-10; creamer, $8-10.

Turquoise Blue cup/saucer, $5-7.

Scarce basketweave bowl, $100-125; straight-sided bowl, $40-45.

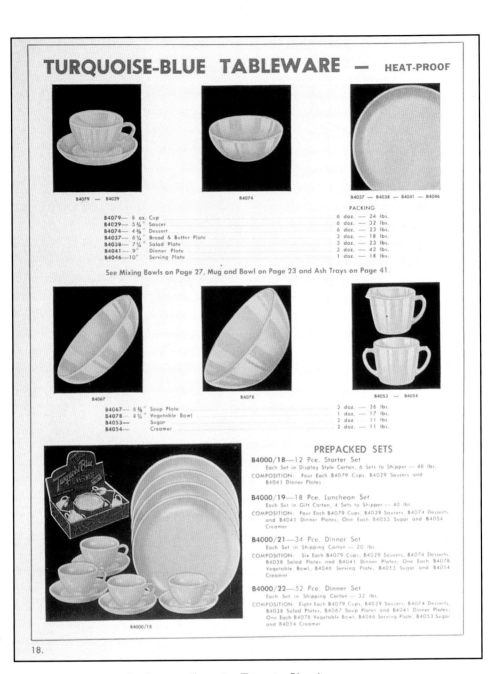

Catalog page illustrating Turquoise Blue dinnerware.

Turquoise Blue snack set with gold trim, $8-10 set.

Turquoise Blue chili bowl, $10-12; D-handled mug, $12-15.

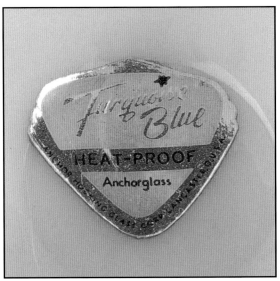

Turquoise Blue label. Adds $2-3 to the value of the piece.

Christmas decoration on Turquoise Blue 6" plate, $25-30.

Fire-King Kitchenware and Miscellaneous Delphite

Three-part relish with gold trim, $10-12, devilled egg tray, $12-15.

Close-up on relish tray label. Adds $3-5 to value.

Breakfast bowls. Azurite with red ivy, $35-40; jadite with red ivy, $100-125; plain jadite, $90-100; plain ivory, $25-30; plain azurite, $40-45.

Azurite chili bowl with advertising metal cover, $20-25.

Fire-King bowls. 6" beaded mixing bowl, $15-20; blue basketweave, $10-12; green basketweave, $12-15.

Three-part Doric candy dish by Jeannette, $8-10.

Doric sherbets. $8-10 each.

Rare Fire-King Sheaves of Wheat plate, $750-1000.

Cherry Blossom tray. $20-25. A small service of delphite Cherry Blossom pieces can be collected.

The back of the Cherry Blossom tray showing the pattern.

Opposite page
Fire-King ashtrays. $20-25 set.

Delphite swirl creamer, $8-10; sugar, $8-10; fruit bowl, $6-8.

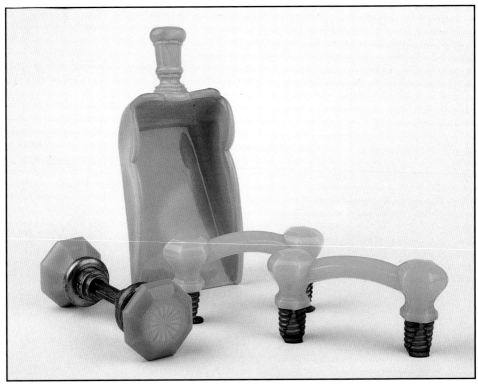

Delphite cologne bottles, $20-25 each.

Opposite page
Top: Chaline eggcups, $20-25.

Bottom: Delphite doorknob, $30-35; scoop, $25-30; Chaline drawer pulls, $45-50 pair.

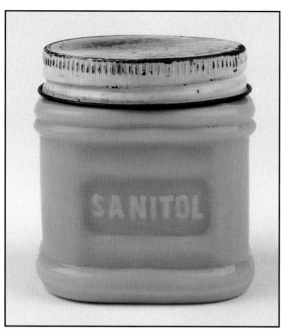

Chaline "Sanitol" jar, $25-30.

Colman's jars. Perfect for toothpicks. $20-25.

Vintage delphite clock, $100-125.

Small Chaline cheese dish, $50-60. Watch for reproductions.

Jadite Kitchenware & Dinnerware

Although Jadite kitchenware has been collected for decades, it is only in the last few years that Jadite dinnerware has become extremely popular. This soft green color, which has been fashionable in recent years, paired with its high visibility in popular magazines and Martha Stewart's TV show, has created an incredible demand for this mass-produced dinnerware.

The canisters and shakers pictured in this book date almost exclusively from the 1930s and 1940s and were made by McKee and Jeannette Glass Companies. Most of the dinnerware was produced by Anchor Hocking under their Fire-King line from 1945 until the late 1960s. For more information, please consult our *Jadite: An Information and Price Guide*. Please be aware that for the last couple of years reproductions and look-alike glassware has been produced. Some of this is poorly made imports. Other pieces are high quality from original moulds. Anchor Hocking has also attempted to reproduce some items.

Opposite page
Top: Jeannette Kitchenware.

Bottom: Jadite mixer, $75-100.

Jadite Kitchenware

McKee 48oz. canisters with press-on lids. $450-500 each. Cereal not pictured.

McKee 48oz. canisters with screw-on lids. Coffee and tea, $325-375 each. Cereal and sugar, $350-400 each. Flour, $375-425.

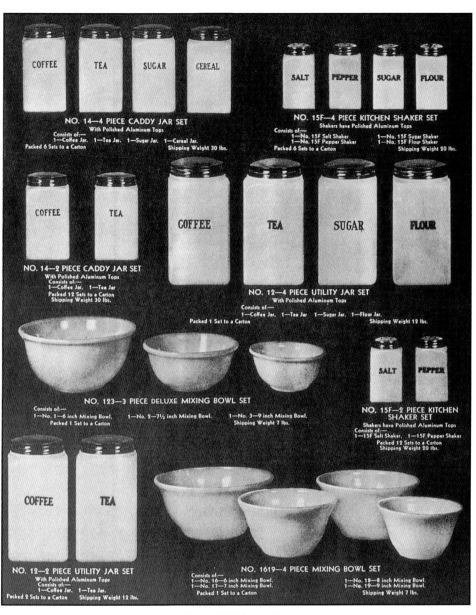

Original McKee catalog.

Comparison of McKee 48oz. and 28oz canisters.

McKee 28oz. Coffee, $300-325.

McKee 28oz. Tea, $300-325.

McKee 28oz. Sugar, $300-325.

McKee 28oz. Flour, $350-375.

McKee 48oz. round canisters with glass lids. $225-250 each.

McKee 28oz. Cereal, $350-375.

Three sizes of McKee round canisters. Lids are interchangeable.

McKee 40oz. tea canister. $200-225.

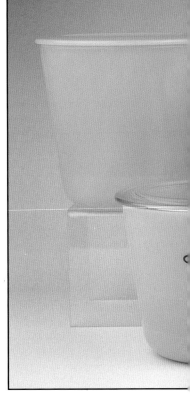

McKee 40oz. canisters. $200-225.

McKee canisters. Canisters with cursive lettering are slightly scarcer. Crystal and black (not shown) lids are also found. Deduct, $25. Canister without lettering, $75-85.

McKee canisters/refrigerator jars. Small, $30-35; medium, $90-100; large, $75-85.

Rare McKee column canisters. Coffee, sugar, flour, $1000-1200 each. Tea, $800-1000.

Jeannette 29oz. sugar canister, $225-250.

Jeannette 29oz. coffee canister, $200-225.

Jeannette 29oz. tea canister, $200-225. Cereal (not pictured), $200-225.

Jeannette 48oz. canisters with Floral lids. Tea, cereal, coffee, $200-225 each. Sugar, $225-250. Flour, $350-400.

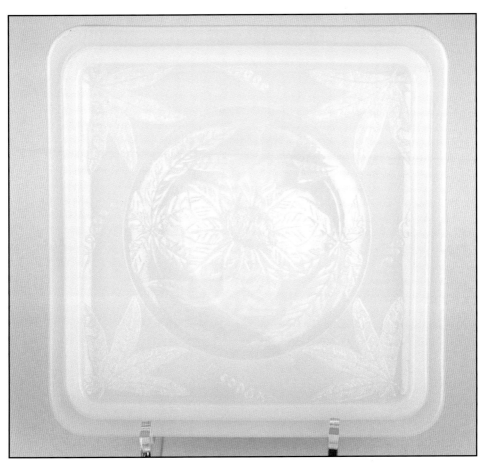

Underside of 48oz. canister lid.

Comparison of Jeannette coffee canisters. Dark canisters are scarcer. Add, $25-75.

Jeannette spice canisters. Nutmeg, allspice, ginger, $150-175; pepper, $125-150; cloves, $500-600.

Children's canisters. Flour, sugar, cocoa, cereal, and tea (not pictured), $450-500. each.

Jeannette canisters. Dark coffee, $250-275; light coffee (not pictured), $225-250; light sugar, $325-350; light tea, $175-200; dark tea (not pictured), $175-200.

Jeannette 40oz. round sugar canisters. Dark, $375-400; light, $325-350.

Fired-on canisters. $200-250 set.

Jeannette 8oz. shakers. Salt, $45-50; pepper, $45-50; sugar, $75-80; flour, $75-80.

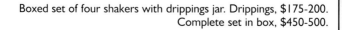

Boxed set of four shakers with drippings jar. Drippings, $175-200. Complete set in box, $450-500.

Jeannette dark 8oz. Salt, $50-60; pepper, $50-60; sugar, $80-85; flour, $80-85.

Jeannette 6oz. shakers with paper labels. $30-35. Without labels, $20-25.

Close-up of "Big Hit" label.

Jeannette 6oz. shakers. Salt, $75-85; pepper, $75-85; flour, $100-125; sugar, $100-125.

McKee Roman Arch shakers. Salt, $85-95; flour, $125-150; sugar, $125-150; pepper, $85-95; spice, $250-275.

Roman Arch shakers with large letters. $75-85 each.

McKee large box shakers. Salt, $60-70; pepper, $60-70; spice, $275-300; sugar, $80-90; flour, $80-90.

McKee small box shakers. Pepper, $60-70; salt, $60-70; flour, $80-90; sugar, $80-90; nutmeg, $175-200; cinnamon, $175-200; spice (not pictured), $175-200; ginger (not pictured), $275-300.

Jeannette sugar shakers with vertical ribs. Light and dark, $125-150. each.

McKee square shakers. Sugar, $80-90; salt, $60-70; flour, $80-90; pepper, $60-70.

Jeannette square shakers. Salt, $60-70; pepper, $60-70; flour, $80-90; sugar, $80-90; allspice, $275-300.

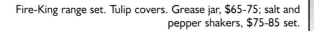

Fire-King range set. Tulip covers. Grease jar, $65-75; salt and pepper shakers, $75-85 set.

Jeannette dark square shakers. Pepper, $65-75; salt, $65-75; sugar, $85-95; flour (not pictured), $85-95.

Ribbed shakers 5-1/4", $100-125 pair.

Jeannette round leftovers. 4-1/4"h x 6"w, $100-125; 3-1/2"h x 6"w, $50-60.

Jeannette leftovers. 4" x 4"$35-40; 4" x 8"$50-60; 5" x 5"with floral lid, $65-75; 5" x 10", $75-85.

Opposite page, bottom: McKee leftovers. 4" x 5", $35-40; 5" x 8"$60-65.

Fire-King Philbe leftovers, $40-45.

5" x 9" loaf pan (plain), $50-60; Philbe 5" x 9" covered leftover, $75-85.

Fire-King crystal covered leftovers. 4" x 4", $35-40; 4" x 8" $55-65.

McKee 4" x 6"leftover (rare), $200-250; clambroth leftover with crystal cover, $30-35.

Fire-King round leftovers. 4-3/4", $100-125; 5-1/2", $60-70.

Fire-King crystal covered butter, $100-125.

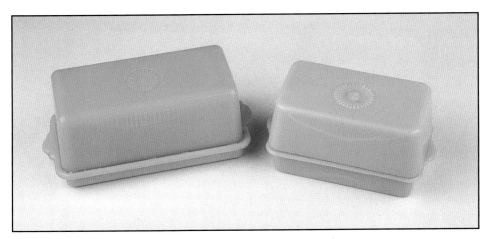

McKee 1-1/4-pounds, $150-175; one-pound, $200-225.

McKee drippings. $450-500 each.

Jeannette round leftover. $35-40.

Jadite-covered crystal leftover, $65-75.

Fire-King mixing bowls.

Fire-King Swirl mixing bowls. 5", $200-250; 6", $35-40; 7", $30-35; 8"$30-35; 9", $35-40.

Fire-King beaded-edge mixing bowls. 4-7/8", $30-35; 6", $25-30; 7", $30-35; 8-3/8", $500-600.

Fire-King Colonial mixing bowls. 6", $125-150; 7-1/4", $125-150; 8-3/4"$125-150.

Fire-King Swedish Modern (teardrop) 5"one-pint, $65-75; 6" one-quart, $175-200; 7-1/4" two-quart, $175-200; 8 3/8" three-quart, $150-175.

Fire-King Splash Proof mixing bowls. One-quart 6-3/4" (rare), $500-600; two-quart, $65-75; three-quart, $125-150; four-quart, $125-150.

Fire-King ribbed bowls. 4-3/4", $100-125; 5-1/2", $60-70; 7-1/2", $50-60.

Jeannette ribbed mixing bowls. Dark bowls are scarcer. Add, $10 each.

Jeanette vertical-ribbed bowl. 6", $35-40; 7", $35-40; 8", $40-45; 9", $75-85.

Jeannette ribbed bowls. 9-3/4", $150-175; 7-1/2", $125-150; 5-1/2", $150-175.

McKee bell-shaped mixing bowls. 6", $30-35; 7", $30-35; 8", $40-45; 9", $55-65.

McKee round bowls. 6", $75-85; 7-1/2", $80-90; 9", $100-110.

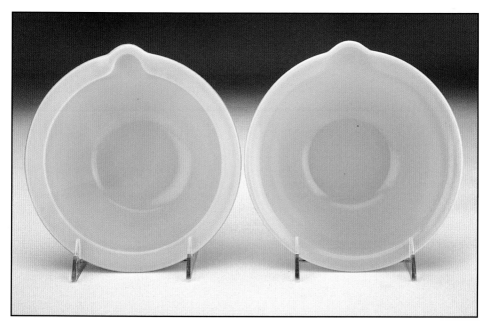

McKee batter bowls. $30-35 each.

McKee 4-3/4" cracker bowl, $75-85.

Fire-King Splash Proof with floral decoration bowl. $500-600 each. Also found with red flowers.

Fire-King skillets. One-spout, $100-125; two-spout, $175-200.

Original large label on skillet. Add $40-50.

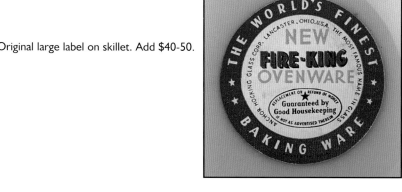

Fire-King pie plates. 9" plain, $275-300; 10-1/2" Philbe, $400-450.

Rear view of Philbe pie plate.

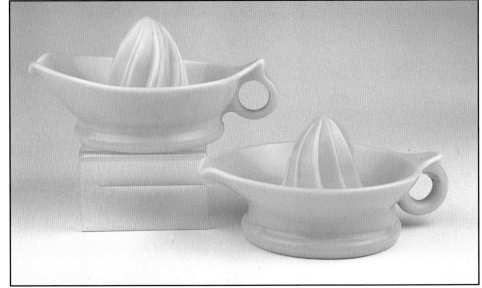

Reamers. Clockwise from left: McKee lemon, $60-70; Jeannette orange, $65-75; Jeanette lemon, $50-55; McKee orange, $70-80; Sunkist, $75-85.

McKee grapefruit. $200-225.

Opposite page, bottom:
Light and dark reamers. Add $5-10 for dark.

Jeannette sunflower pitchers with reamer top. Light, $125-135; dark, $150-160.

Fire-King pitchers. Bead and Bar milk pitcher, $275-300; plain milk, $85-95.

Jeannette sunflower pitcher. $100-125.

Jeannette ice box jug with lid, $500-600.

Fire-King batter pitchers. Colonial 1" band, $65-75; 3/4" band, $60-70.

Large label on batter pitcher. Adds $10.

Original ad for batter pitcher.

McKee water servers. Round lid, $450-500; tall water server with rectangular lid, $600-650.

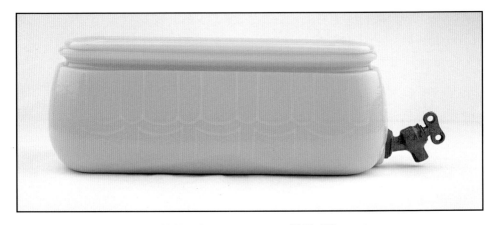

McKee short water server, $350-400.

Clambroth water server, $150-175.

Fire-King Plain ball jug, $650-$750.

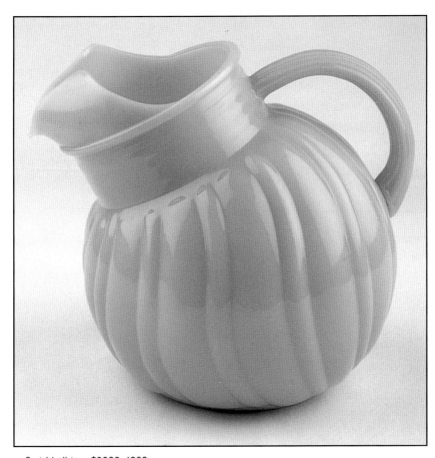

Swirl ball jug, $3000-4000.

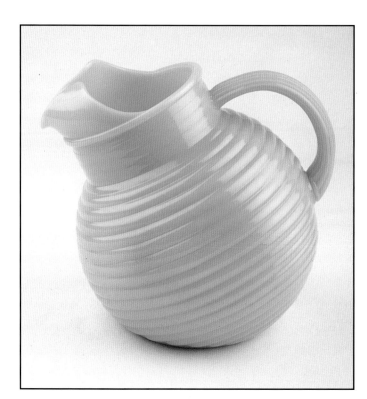

Manhattan ball jug,
$3500-4000.

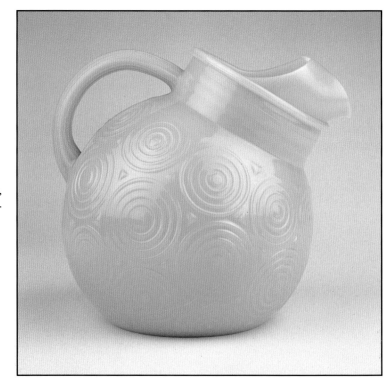

Target ball jug,
$4000-5000.

McKee handleless one-quart, $1200-1500.

View of rolling pin's shaker top.

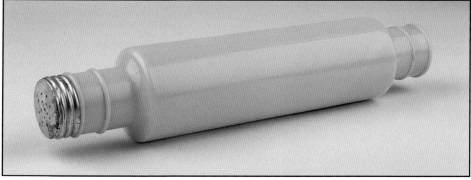

McKee rolling pin. $700-800. Watch for Martha Stewart's reproduction, which are very close to the original and retail for around, $65.

McKee vases. From left. 6-5/8" bud vase, $100-125; 8" ribbed, $125-150; 4" vase, $100-125; 10-1/2" vase, $150-175; 6-1/2" bulbous vase, $150-175.

Opposite page: McKee large swing vase, $250-300.

Jeannette measuring cups. One-cup, $60-65; half-cup, $45-50; third-cup, $40-45; quarter-cup, $30-35. Complete set in box, $200-225.

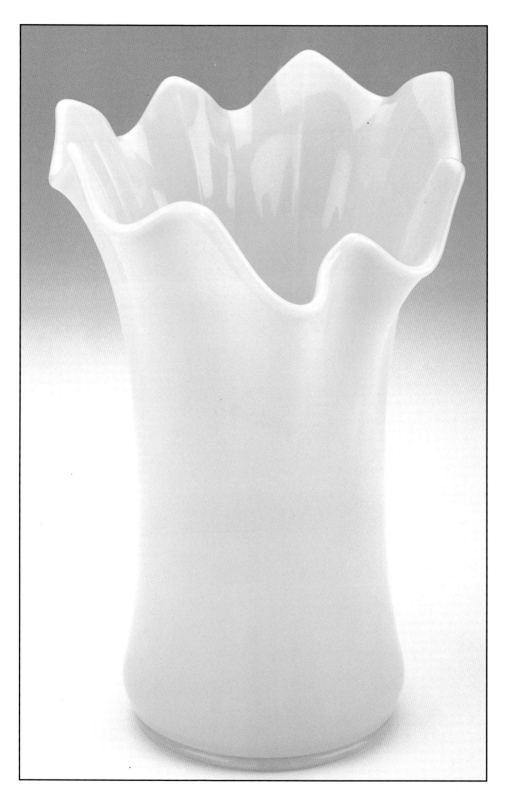

McKee 11-1/2" vase, $200-225; 8" vase, $150-175.

McKee "nudes." "Art Dressed" (scarce), $250-300; 7" Art Nude, $200-225 (with cover, $275-300); 8" Art Nude, $200-225.

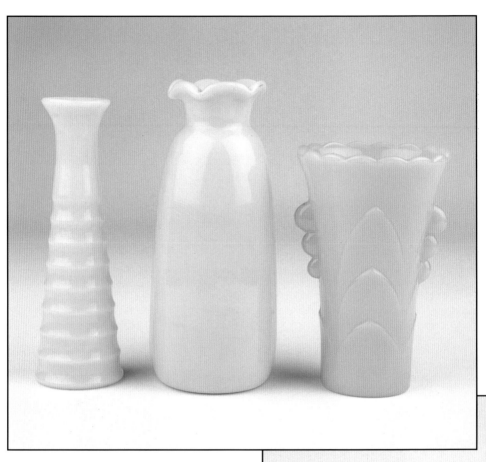

Jeannette bud vase, $25-30; McKee 6-5/8" bud, $100-125; Fire-King Art Deco, $20-25.

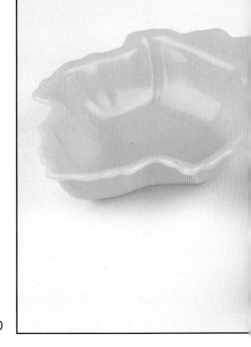

Fire-King dishes. Leaf, $20-25; shell, $25-30; comport, $75-85.

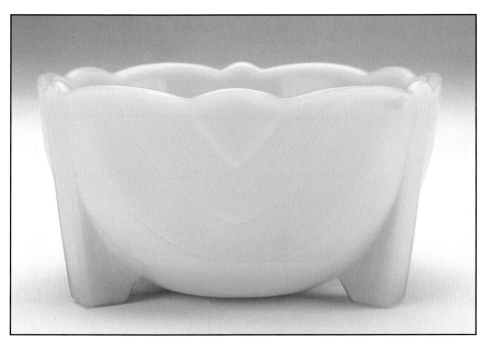

McKee bulb bowl. $45-50.

McKee lion planter, $100-125.

Fire-King flower pots. 3-5/8" smooth edge, $20-25; 3-1/4" ruffled, $25-30.

GIFT PIECES IN BEAUTIFUL JADE-ITE

G846—4¾" Dessert Cup
Pkd. 4 doz. ctn.—wt. 20 lbs.

G847—8" Salad Plate
Pkd. 4 doz. ctn.—wt. 38 lbs.

G807—3½" Flower Pot
Pkd. 4 doz. ctn.—wt. 25 lbs.

G810—5" Vase
Pkd. 4 doz. ctn.—wt. 30 lbs.

G820—5¼" Bulb Bowl
Pkd. 4 doz. ctn.—wt. 33 lbs.

G388—8½" Fancy Bowl
Pkd. 2 doz. ctn.—wt. 30 lbs.

G997—4½" Cigarette Box
& Cover
Pkd. 2 doz. ctn.—wt. 22 lbs.

G822—4¼" Ash Tray
Pkd. 4 doz. ctn.—wt. 19 lbs.

Anchor Hocking Glass Corporation
LANCASTER, OHIO, U. S. A.

Fire-King catalog page.

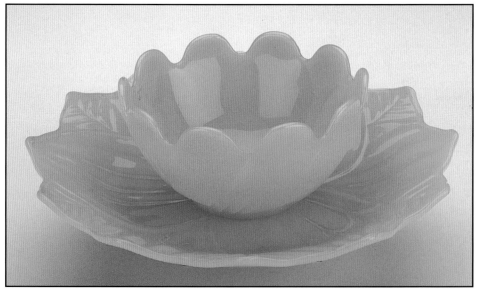

Fire-King leaf and blossom. Leaf plate, $20-25; blossom bowl, $20-25.

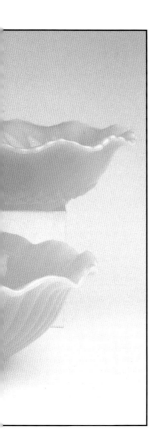

Fire-King bowls. Clockwise from left: 6" ribbed-bulb bowl, $30-35; diamond-bottom bowl, $35-40; ruffled-diamond bowl, $70-80; 7" swirled-bowl with ruffled edge. $45-50.

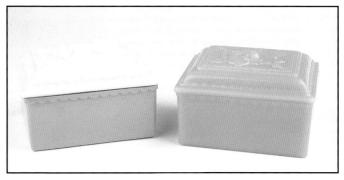

Fired-on jewel box, $25-30; Jadite jewel box with rose cover, $75-85.

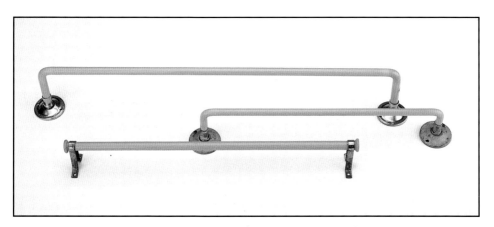

Towel bars. $25-30 each.

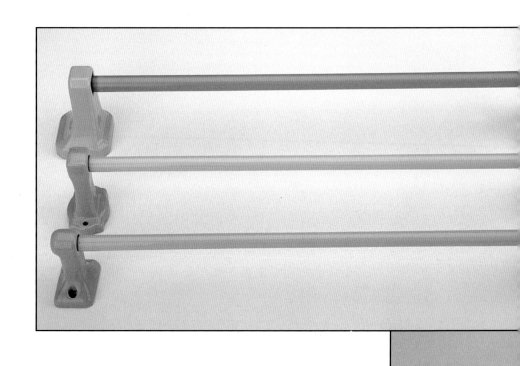

Fire-King Alice cup/saucer, $10-12, dinner, $35-40.

Towel bars with ceramic holders, $35-40 each.

Alice

Charm

Charm sugar, $20-25; creamer, $20-25.

Charm bowls. 4-3/4" dessert, $18-20; 6" soup, $50-60; 7-3/8" salad, $75-85.

Charm cup/saucer, $12-15.

Opposite page
Top: Fire-King Charm dinnerware. Platter, $70-80.
Bottom: Charm plates. 6-5/8" salad, $35-40; 8-3/8" lunch, $20-25; 9-1/4" dinner, $60-65.

Jane Ray

Fire-King Jane Ray dinnerware.

Jane Ray plates. 9-1/8" dinner, $14-16, 7-3/4" salad, $12-15; 6-1/4" bread plate, $100-125.

Jane Ray platter, $30-35.

Jane Ray bowls. 8-1/4" salad, $35-40; 7-5/8" soup, $25-28, 4-7/8" dessert, $10-12, 5-7/8" cereal, $20-25.

Jane Ray flanged soup, $500-600.

Jane Ray cups/saucers. Regular, $10-12; demitasse, $75-85.

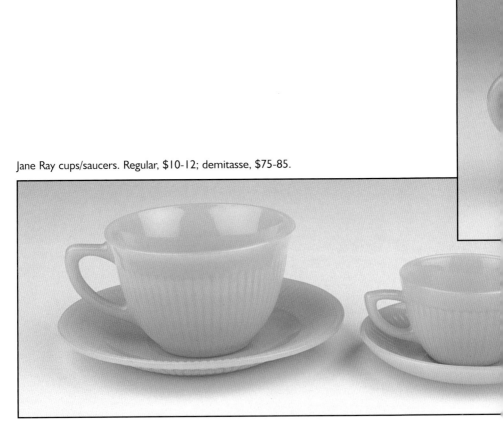

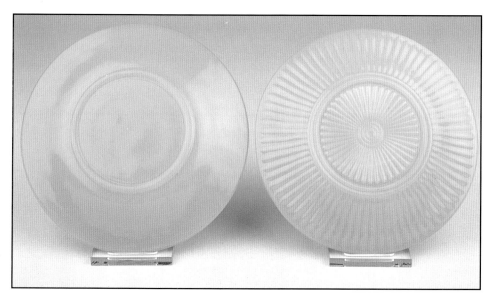

Two styles of Jane Ray saucers.

Jane Ray creamer, $18-20; sugar, $40-45.

McKee Laurel

McKee Laurel plates. 9" dinner, $20-25; 7-1/2" salad, $25-30; 6" bread, $10-12.

Laurel plates come with a scalloped or plain edge.

Laurel 9" grill, $20-25; platter, $35-40.

Laurel cup/saucer, $15-18.

Laurel bowls. Clockwise from left: 11" round utility, $75-85; 9" round vegetable, $35-40; 9" oval vegetable, $45-50; three-footed jelly, $30-35; 6" cereal with flanged rim, $20-25; 6" oatmeal, $30-35; 5" fruit, $12-15.

Laurel sugar, $20-25; creamer, $20-25.

Laurel salt and pepper, $75-100 pair.

Laurel cheese dish, $250-300.

Laurel candles, $85-95.

Laurel children's pieces with Scottie dog. Cup/saucer, $125-150; plate, $50-60; creamer, $125-150.

Restaurantware

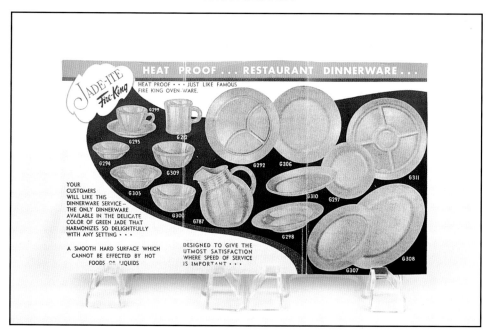

Fire-King Restaurantware brochure.

Restaurantware plates. 9" dinner, $22-25; 8" lunch, $50-55; 6-3/4" pie, $10-12, 5-1/2" bread, $12-15.

Restaurantware bowls. Clockwise from top. Flanged soup, $100-125; flanged cereal, $25-30; chili bowl, $18-20; 10oz. bowl, $35-40; fruit bowl, $10-12, 15oz. bowl, $35-40.

Restaurantware mugs. Three styles/sizes, $25-28 each.

Restaurantware cups and saucers. From left: tall cup/saucer, $18-20; regular cup/saucer, $15-18; demitasse, $75-85; tapered cup/saucer, $30-35.

Restaurantware platters. Clockwise from top: 9-1/2" football, $75-75; 8-7/8" oval, $100-125; 8-7/8" divided, $85-95; 9-1/2" oval, $50-60; 11-1/2" oval, $50-60.

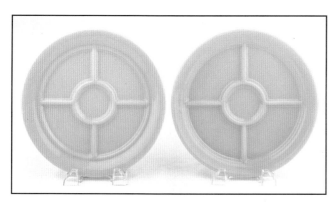

Restaurantware five-part plates. $45-50 each.

Restaurantware grills. $25-30. each.

1700 Line

1700 line and breakfast set. Plate, $25-30; eggcup, $40-45; breakfast bowl, $90-100; St. Denis cup/saucer, $12-15; milk pitcher, $90-100.

Philbe mug with design, $125-150; "Wannabe" mug, $20-25.

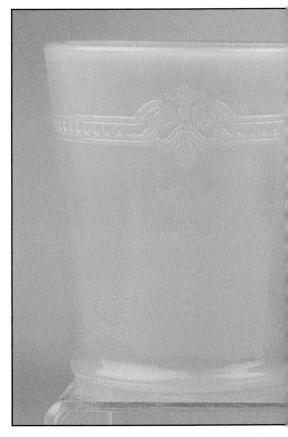

Ransom cup
saucer, $20-22;
St. Denis cup/
saucer, $12-15.

Sheaves of Wheat

Close-up on Sheaves of Wheat plate.

Fire-King Sheaves of Wheat. Cup and saucer, $100-125; plate, $80-90; bowl, $125-150.

Shell

Fire-King Shell dinnerware. Platter, $75-85; 7-1/4" salad plate, $15-20.

Shell dinner, $25-30.

Shell cup/saucer, $12-15.

Shell bowls. 8" vegetable, $30-35; 7-5/8" soup, $40-45; 6-3/8" cereal, $20-25; 4-3/4" dessert, $12-15.

Shell sugar bowl, $20-25; cover, $50-60; creamer, $20-25.

Swirl

Fire-King Swirl cup/saucer, $100-125; dinner, $90-100.

Martha Stewart

Martha Stewart cake plates.

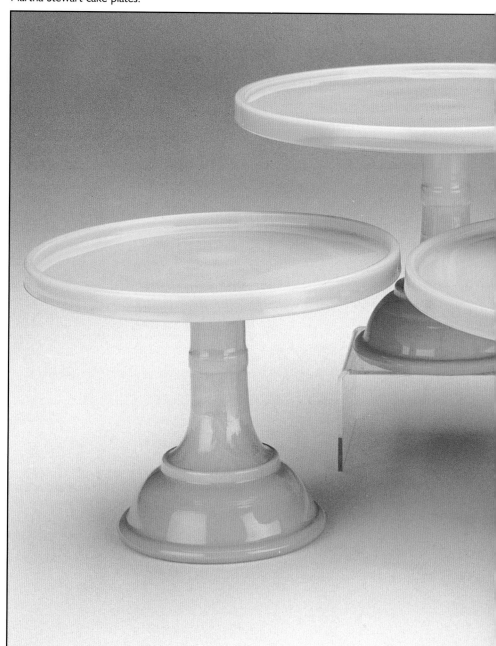

Martha Stewart cake plate.

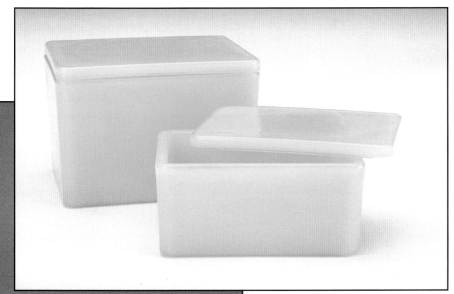

Martha Stewart covered boxes.

Martha Stewart items.

Martha Stewart tumblers.